CAREERS IN THE

UNITED STATES COAST GUARD

CHOOSING A CAREER SHOULD BE AN interesting, eye-opening experience. You are entering a time when you will have to make big decisions about the future direction of your life. You owe it to yourself to get them right.

The fact that you have chosen to read this report is a good sign that you are headed in the right direction. It is also an indication that you want a career defined by challenge, service and adventure. You will definitely find it with the United States Coast Guard, America's "fifth armed service" and the lead agency in the Department of Homeland Security. The Coast Guard is one of the hardest-working federal agencies, and one that is held in very high esteem by the American people it serves. Coasties get respect.

The Coast Guard is the fifth largest federal armed service, after the Army, Navy, Air Force and Marine Corps. The Coast Guard is only about the same size as the New York City Police Department – both services have about 40,000 full-time personnel. In recent years, the NYPD has actually become smaller because crime in New York City has decreased. The Coast Guard has grown larger because the kind of crime it deals with has increased.

The Coast Guard is a military organization, however it devotes most of its efforts to conducting law enforcement missions. The Coast Guard is unique in this regard because the other military services are prohibited by law from engaging in law enforcement. Operating under Title 10 of the United States Code, the Coast Guard can enforce laws, as well as conduct military operations for the Department of Defense or directly under the command of the President. The Coast Guard also maintains a robust search-and-rescue capability, protects the marine environment, and provides aids to navigation. It is also the lead agency in the Department of Homeland Security, the federal department created after the terrorist attacks of September 11, 2001. Coasties get respect because they work very hard and have many missions that are vital to the safety and security of the United States.

If you like what you read in this report, your next stop should be the nearest Coast Guard recruiting station. The service's needs change constantly, and only a recruiter can give you the latest scoop on what kind of jobs are available and where. Remember, by just talking to a recruiter you incur no obligation. You are not committed to join the Coast Guard until you sign a contract. So visit a recruiting station, pick up a few brochures, and chat with the experts.

WHAT TO DO NEXT

DO NOT PUT OFF PREPARATION FOR YOUR Coast Guard career until you report to boot camp. There are several steps you can take now.

First, concentrate all your efforts on doing well in school. Coasties wear multiple hats. Enlisted personnel will be assigned to jobs (called rates) in which they will receive specialized training. Officers specialize but are generally required to serve in several "out-of-specialty" tours over the course of a career, an arrangement that requires Coast Guard officers to be capable generalists. Unlike the other military services, which tend to train constantly while preparing for the next war, the Coast Guard fights a low-level war every day of the year. In this respect the Coast Guard is much more like a police department than a military service. This constant operational tempo, known as the OPTEMPO in military lingo, added to the Coast Guard's small size, makes it necessary for everybody to be prepared to do everything. So study hard while you are still in high school. You will not get very far in the Coast Guard without a solid educational background.

Coasties get closer to the practicalities of maritime operations than most of their Navy counterparts do. Ships are smaller, they spend more time underway, personnel are expected to pitch in with everything required, and many of the Coast Guard's primary functions, like search and rescue, are all about getting wet. Learn the basics of boating. Many organizations offer classes in boating, including the Coast Guard, United States Power Squadrons, Safe Boating America and many park districts and yacht clubs. You do not have to become a skilled sailor to succeed in a career in the Coast Guard, but you should be comfortable around small boats and not prone to seasickness. You should also be a strong swimmer. Lifesaving certification is essential.

The Coast Guard is a legendary institution, so there is no shortage of material available to help you learn more about it. Notable movies include *The Guardian; Yours, Mine & Ours; The Perfect Storm; The Boatniks; The Fighting Coast Guard,* and the all-time classic *The Beast from 20,000 Fathoms.* The privately owned newspaper *Navy Times* covers the Coast Guard, and the service publishes its own magazine, *Coast Guard Magazine.* There are numerous books available too, including *Character in Action: The US Coast Guard on Leadership* by Donald T. Phillips and former Coast Guard Commandant James M. Loy.

HISTORY OF THE CAREER

IN ITS CURRENT FORM, THE UNITED States Coast Guard is a very young organization, founded in 1915. Its roots, however, reach all the way back to 1790 and the creation of the Revenue Cutter Service. A 10-ship maritime service founded by Congress to enforce trade regulations along the American coast, the Revenue Cutter Service also served as the new country's only armed maritime service from 1790 to 1798. The mighty American navy, which would grow to many times the size of the Coast Guard, existed only on paper between the end of the Revolutionary War and 1798, when the Continental Navy formally became the US Navy. During those years the Revenue Cutter Service was all that stood between the fledgling United States and often-belligerent European powers.

The Revenue Cutter Service took on important jobs right from the beginning. In 1794, it was tasked with preventing slave ships from Africa from landing in the United States. In 1798 and 1799, it participated in the so-called Quasi War with France. In the early 19th century, President Thomas Jefferson closed all US ports to European ships, an order enforced by the Revenue Cutter Service. The Service also took the lead in enforcing one of the nation's first environmental laws, the 1822 Timber Act, which aimed to prevent illegal logging on American territory. Revenue cutters were assigned to blockade duty during the Civil War and went on to play a pivotal role in opening up Alaska to settlement and commerce late in the 19th century.

The modern Coast Guard was created in 1915, when the Revenue Cutter Service was combined with the United States Life Saving Service, another small federal agency. It was not the last time the Coast Guard would be augmented by taking on additional functions. In 1939, the Lighthouse Service was merged into the Coast Guard, and in 1942, the Bureau of Marine Inspection and Navigation, which had already absorbed the Steamboat Inspection Service, was folded into the Coast Guard. The new agency remained part of the Department of the Treasury, where the Revenue Cutter Service had been founded in 1790.

Heralding the Coast Guard's eventual top-to-bottom retooling as the lead agency in the Department of Homeland Security, the various mergers that created the modern Coast Guard illustrate both the Coast Guard's importance and its unique place among military services and law-enforcement agencies. The Coast Guard's mission may seem obvious: to guard the coast. At least five different federal agencies were originally created to take on various aspects of this mission, from running lighthouses to inspecting steamboats, and they all are now part of the Coast Guard. The US Navy is responsible for maritime defense far from home, but the Coast Guard can be seconded to the Navy in time of war. As you can see, the scope of the Coast Guard's mission is enormous and quite varied.

The zeal to persevere is immortalized in the Coast Guard regulations adopted from the Life Saving Service. When attempting a rescue, the regulations say, all options must be tried before breaking off the attempt:

"In attempting a rescue the keeper will select either the boat, breeches buoy, or life car, as in his judgment is best suited to effectively cope with the existing conditions. If the device first selected fails after such trial as

satisfies him that no further attempt with it is feasible, he will resort to one of the others, and if that fails, then to the remaining one, and he will not desist from his efforts until by actual trial the impossibility of affecting a rescue is demonstrated. The statement of the keeper that he did not try to use the boat because the sea or surf was too heavy will not be accepted unless attempts to launch it were actually made and failed, or unless the conformation of the coast – as bluffs, precipitous banks, etc. – is such as to unquestionably preclude the use of a boat."

In the run-up to World War II President Franklin Delano Roosevelt ordered the Coast Guard to patrol the North Atlantic under the pretext of keeping an eye on icebergs. During the war, the Coast Guard routinely rescued Americans and allies whose ships had been torpedoed by German submarines prowling the important shipping lanes between the United States and the United Kingdom, the island that would later serve as the launching point for the Allied invasion of Europe. The Coast Guard supported the Navy throughout World War II. It played an important role in the D-Day invasion of Europe in 1944 by deploying a fleet of 60 cutters to handle search-and-rescue operations during the invasion. The Coast Guard managed to sink a dozen German submarines during the war, a testament to the service's willingness to do whatever is necessary to get the job done.

The Coast Guard participated in the conflicts in Korea and Vietnam, supporting the Navy with its expertise in search and rescue, and operations near the shore. This was in addition to responding to the steady stream of rescues and law-enforcement activities in American waters. In 1967, the Coast Guard was transferred from the Treasury Department to the Transportation Department, a move thought to better reflect the service's emphasis on domestic law-enforcement in the maritime commons. Also in 1967, the Coast Guard adopted the "racing stripe" insignia for use on its cutters. The red-and-blue stripe proved so popular many international coast guards adopted it as a universal means to tell a coast guard vessel from a navy vessel at a glance.

The Coast Guard's world changed dramatically from the 1970s to the 1990s. Women were fully integrated into the service in 1978, resulting in a significant cultural shift in a service that historically prided itself on its "wooden ships and men of steel." Enhanced environmental awareness brought on greater responsibility for environmental protection, especially in preventing illegal fishing. The United Nations Convention on the Law of the Sea established new formal boundaries for nations bordering on the sea – 12 miles of territorial water and 200 miles of Exclusive Economic Zone – and changed the way the Coast Guard patrols its home waters.

The biggest change to come to the Coast Guard in recent years was its transfer to the Department of Homeland Security in 2002. Created in response to the terrorist attacks on the United States on September 11, 2001, the Department of Homeland Security's goal is to prepare for, prevent and respond to domestic emergencies, and especially acts of terrorism. The new department swallowed up dozens of smaller federal agencies controlling border patrol, customs, and immigration, and quickly became the third largest Cabinet level department.

The urgency created by the terrorist attacks and the creation of a new agency put the Coast Guard's unique combination of military prowess and law-enforcement expertise squarely in the public eye. As a result, the Coast Guard's budget was substantially increased, allowing the service to add approximately 5,000 new enlisted and commissioned members and make progress on its long-discussed Deep-water program. Intended to replace all of the Coast Guard's ships and aircraft, the Deep-water program's main objective is to make sure that the Coast Guard sails into the future with a balanced fleet equipped with systems that are interoperable throughout the organization and with the Navy.

The future of the US Coast Guard is very promising, and you can be a part of it.

WHERE YOU WILL WORK

MILITARY PERSONNEL TEND TO MOVE often. Career paths require that personnel get different experiences in different places. The Coast Guard is also a continuum of people; that is, there are always people joining at one end and getting out or retiring at the other end. Billets (work assignments) still need to be filled, but the people filling them today will move along by tomorrow. If you pursue a career with the Coast Guard you can expect to move every three years or so.

Of course, much of your work will take place aboard Coast Guard vessels, known as cutters. Even though the largest cutters, known as high-endurance cutters, are only as large as a medium-size Navy ship, you will find that they are fairly comfortable. The food is good and there is nothing quite like being gently rocked to sleep by the motion of the waves. It is easy to get used to. You will have a bed, known as a rack, a small closet and a few drawers to call your own. It is not much but you will find that it is all you need. When you're on sea duty, you're on duty, not a pleasure cruise.

Cutters are home ported at Coast Guard installations around the United States. The service's single largest installation is in Kodiak, Alaska. A small island, Kodiak is home to a large commercial fishing fleet and is well positioned to allow cutters to patrol a wide swath of very dangerous waters. Statistically speaking, commercial fishing is among the most dangerous professions in the world. Fishing boat crews like working near Coast Guard bases for the safety and protection they can be called on to provide. Coast Guard district headquarters installations are located in the coastal cities of Boston; Portsmouth, Virginia; Miami; New Orleans; Alameda, California; Seattle; Juneau, Alaska; and Honolulu, Hawaii. The Ninth Coast Guard District is headquartered on Lake Erie in Cleveland, Ohio and serves the Great Lakes.

The Coast Guard also maintains small installations and piers on the East, West and Gulf Coasts and around the Great Lakes. The service often deploys small boats to inland waterways to assist local authorities with patrolling during busy seasons. Recruiting stations are located in most major metropolitan areas. If you stay in for a few hitches, you may find yourself on recruiting duty a long way from a major body of water.

The Coast Guard does not maintain any permanent overseas facilities. It does, however, deploy individual Coast Guard personnel and Coast Guard units to other US military facilities around the world. It is very common for Coast Guard cutters to deploy with US Navy battle groups, bringing their special capabilities to bear in support of the Navy's blue-water expertise. A Coast Guard career can take you anywhere.

DESCRIPTION OF YOUR WORK DUTIES

COAST GUARD OFFICERS AND ENLISTED personnel have similar career paths. For example, enlisted boatswain's mates will always be working for officers assigned to operational billets afloat. Both specialties are concerned with piloting Coast Guard vessels, or driving ships in military lingo. Officers lead the boatswain's mates and other enlisted shipboard personnel. The enlisted personnel do the actual work of keeping the ship afloat. It is a very efficient system and one that gives individual careerists opportunities to play to their strengths and move up in the world. There is no question that officers have more prestige; they are the leaders, they have more authority and they receive higher salaries. Enlisted personnel are the Coast Guard's technical experts. Do not get caught up in the notion that it is always more desirable to be an officer. If you want to lead people and manage projects, apply for a commission. If you want to become an expert rescue swimmer or intelligence specialist – or have any number of other rewarding jobs – enlist. Take the route that is better for you.

Enlisted Jobs

Enlisted job classifications are divided into groups and ratings. Groups are skill clusters that contain several related jobs, known as ratings. Demand for new accessions rises and falls all the time. Some ratings can become full, making advancement difficult. Others may have a shortage of skilled Coasties, making it easy to fill billets. This is why you need to talk to a recruiter to get the latest details on job opportunities. Some ratings may be temporarily closed to new accessions while others may be desperate for new recruits. You need to know the score in order to make the best choice for your career goals.

The Deck and Ordnance Group contains jobs related to operating Coast Guard vessels. Largest of these ratings is boatswain's mate (pronounced "bosun"), the practical hands-on sailors who keep cutters running. Gunner's mates also fall into this group. Gunner's mates are responsible for all small arms and shipboard weapons systems. Operations specialists are responsible for command, control and communications. Operations specialists work mainly in critical operational areas, like the bridge and combat information center. Intelligence specialist is a new rating for the Coast Guard. Intelligence specialists are responsible for gathering, analyzing and disseminating intelligence critical to meeting the Coast Guard's many missions. Intelligence specialists have to undergo rigorous background checks to make sure they can be trusted with classified data.

Ratings in the Hull and Engineering Group are concerned with the maintenance of Coast Guard vessels. This is where the heavy-duty mechanical jobs can be found. A modern Coast Guard cutter is a very sophisticated piece of equipment. If you like to work with your hands on machines, here is where to look. Damage control men specialize in keeping the ship afloat and operational after it has been damaged in combat or as a result of a mishap. Electrician's mates keep the cutter's complex electrical systems up and running. Electronics technicians maintain the electronic instruments and systems aboard the vessel, including all command, control and computer systems. Information systems technicians maintain the networks that link the cutter departments to each other and the cutter to the rest of the world. Machinery technicians work on the cutter's engines and other major machinery.

Ratings in the Aviation Group keep the Coast Guard in the air. The Coast Guard maintains one of the world's largest fleets of aircraft, both fixed-wing (regular airplanes) and rotary-wing (helicopters). They are all maintenance-intensive and offer very small margins of error. Avionics electrical technicians maintain the high-tech flight systems common to all modern aircraft. Aviation maintenance technicians maintain the aircraft themselves, from engines to fuselages.

Aviation survival technicians are the Coast Guard's legendary rescue swimmers. They jump out of helicopters to rescue people stranded in the water. They also have the paramedic skills necessary to administer first aid at the scene. This is the kind of job that movies are made about. It is extremely difficult work that can lead to heroic efforts and dramatic lifesaving feats.

The Administrative and Scientific Group includes all of the functions necessary to support a large organization like the Coast Guard. Food service specialists are responsible for preparing meals both in shore-based facilities and in galleys aboard cutters. Many food service specialists say that the need to prepare many meals in tiny cutter galleys is the best professional cooking training they could ever get. Health services technicians provide support services in healthcare facilities and in the field. Marine science technicians are critical to the Coast Guard's environmental protection mission. They test water, conduct environmental surveys and monitor environmental challenges in American waters. Public affairs specialists keep the Coast Guard in touch with American citizens and public opinion throughout the world. Modern public affairs training covers photography, videography, writing, crisis communications and all manner of digital media tools. Storekeepers keep track of supplies and are a critical link in the Coast Guard's logistical chain.

Yeomen handle paperwork and records. Although the smallest of the armed services, the Coast Guard is still an armed service and, therefore, prone to generating an enormous amount of paperwork. Everybody loves a good yeoman.

The Work of Officers

Officer career paths cover the same basic functional areas as those of enlisted personnel. All chains of command have to end with officers. Coast Guard officers have a specialty just like enlisted personnel but are also expected to be generalists. This makes the Coast Guard unique among the armed services. All services expect officers to be capable leaders and administrators in addition to being good at their specific jobs. However, no other service specifically requires officers to do an occasional tour outside their specialty. The Navy, for example, will never put an intelligence officer in a human resources billet. The Coast Guard makes such seemingly oddball assignments all the time. The smallest service has been chronically understaffed and underfunded for so long that its corporate culture has evolved to make the best of a challenging situation. Coast Guard officers have to be able to do everything. Most Coast Guard officers would not have it any other way.

Aviation officers are the Coast Guard's pilots. They fly fixed- and rotary-wing aircraft ranging from lumbering C-130 cargo planes to nimble rescue helicopters. Command, Control, Communications and Computers officers, often known as C4 officers, are responsible for the Coast Guard's array of electronic communications and computing equipment. Engineering logistics officers keep track of engineering programs and the supplies and materials that make them go. Civil engineering officers tend to the Coast Guard's physical facilities on shore. Naval engineering officers participate in the design and maintenance of sophisticated Coast Guard vessels.

Financial resource management officers keep track of the money, while human resource officers keep track of the people. Both of these are very big jobs. Health services officers are the doctors and nurses who keep Coast Guard personnel healthy so they can do their jobs. Legal officers are lawyers who enforce the Uniform Code of Military Justice and serve as legal advisors to the service. Marine safety officers specialize in the Coast Guard's mission to place and maintain aids to navigation. Operations officers can be assigned to afloat, ashore, intelligence or law-enforcement billets. This specialty includes ship drivers. Officers assigned to reserve program management run the Coast Guard's part-time reserve program.

Most officer career paths come with hands-on training. Some, like legal, medical and engineering career paths, require degrees in specific subjects in order to be eligible for a commission. All officer careers require at least a bachelor's degree, and most will require that you earn an advanced degree as you move up in the chain of command.

TRUE TALES FROM THE COAST GUARD

I Am a Rescue Swimmer

"Officially I'm known as an aviation survival technician. In the Coast Guard we call ourselves rescue swimmers. We've always been rescue swimmers; we always will. When I see "aviation survival technician" on my paperwork I have to remind myself that that is me.

Anyway, I got into this career because I couldn't think of anything more exciting. Flying into storms to rescue people in distress sounded like a great way to make a living. I got my start the same way most rescue swimmers do, by being on my high school swim team. Aviation survival technician is not a rating you can just apply for. You have to pass a very rigorous swimming test in order to be selected for training, and I mean rigorous. To pass it, you'd pretty much have to be a high-school swimmer, and one with a shelf full of trophies. Everybody I know started out this way.

All those high-school stars make the rescue swimmer training very competitive. Just because you were hot stuff in your hometown doesn't necessarily mean you are even half as good as somebody from a completely different town a thousand miles away. The military brings together people from all corners of the United States. It's one of the things I like best about this job. It was hard during training, though. Even after passing the challenging tryout, about 30 percent of rescue swimmer trainees drop out of the program. Most of them go on to other careers in the Coast Guard.

In an average day, the Coast Guard saves 15 lives and rescues another 114 people in distress. Rescue swimmers account for most of these stats. I can't kid you; this is a really dangerous way to make a living. Think about it for a second. Ships and pleasure craft are most often at risk when they get caught in bad weather. When we get a distress call we have to go into a storm that all sane people are trying to get out of – and we go in a helicopter! Then we have to jump out of the helicopter into the stormy water. It's scary, every time. But we know that we have the best training in the world. We also know that people will die if we don't jump out of the helicopter and do our best. So out we go.

I love this career for many reasons. Like most military professionals, I like being part of something larger than myself. I also like the Coast Guard culture, which is very self-reliant. What I like most, though, is saving lives. There is no substitute for the feeling you get when you return to shore with a helicopter full of people happy to be alive. Oh, I put on a brave face and say things like, 'Just doing my job,' but I still get kind of emotional about what I've done."

I Am a Helicopter Pilot

"Sometimes I think I must be nuts. I fly helicopters into storms – on purpose. We lose a few helicopters and crews ever year but this is still one of the most sought-after officer billets in the Coast Guard. I guess it's the same reason the enlisted rescue swimmer program always gets more applicants than it can possibly use. We live life to the limit, every day.

I majored in history in college. I thought maybe I would

join the military someday, and go into intelligence. My school did not have an ROTC program so I wasn't directly exposed to the military. I earned a private pilot's license while I was in college because I had a hunch it might be valuable later in life. I was right.

I joined the Coast Guard because I really like the scope of the mission. No other service does so much with so little. I also like the emphasis on being a generalist. After I passed the flight physical and was told that I would qualify for flight training I had to make a choice: fixed-wing or rotary-wing. I already had some fixed-wing experience and rotary-wing had always seemed to me to be the essence of what the Coast Guard is all about.

I fly the HH-65 Dauphin helicopter, the Coast Guard's short-range search-and-rescue helicopter. The helicopter is made by a French company and has an American-made engine. It is utterly purpose-built. It is made from composite materials to be light and stay free of rust, which is a major consideration for anything that will spend most of its life around water. It is powerful and very maneuverable. It can fly in conditions that would eat other helicopters alive.

My typical day will be familiar to anybody in law enforcement. Mostly, I sit in a ready room with my co-pilot and a couple of rescue swimmers. We watch a lot of movies. We wait for the call from our emergency dispatch center. When we get one, we have a ritual. Everybody goes to the bathroom and we all get on the helicopter. Then we're off.

We don't always fly into storms. Accidents can happen for all sorts of reasons. Sometimes all we have to do is pluck people out of calm waters because they did something stupid, not because the weather was bad. That's an easy day.

A hard day starts with an emergency call during a nasty storm – usually at night. At the far end of the helicopter's range. That's a hard day. We can't always find the people we're looking for right away. Their distress beacons may stop working, or they may not have one and were called in by a passing aircraft. Darkness makes this process harder. When we do find who we're looking for we have to come in low enough to deploy rescue swimmers, which can be tricky. We have lost helicopters to waves because pilots came in too low. Swimmers have to find the people who need to be rescued and then get them into the steel basket we lower for them. This is easier said than done. Sometimes they are injured and have to be carried into the basket. Often they are panicky and have to be wrestled away. I've encountered a few who just didn't want to leave their multimillion-dollar boats behind.

For a helicopter pilot, it does not get any better than this. This is the world's greatest, most daring and highest-profile helicopter-flying job in the world. I wouldn't trade it for anything. If you are thinking about getting into this career, I have to warn you, you need to be a little bit crazy."

I Am a Naval Engineering Officer

"I always knew I wanted to build ships. I grew up near the water and was an accomplished sailor by my teens. Some of the first adults I got to know when I was a kid were members of the Coast Guard Auxiliary, who performed boat safety checks at the recreational marinas where I hung out. I thought they had pretty cool jobs.

I majored in naval engineering in college, which is a necessity if you want to apply for a Coast Guard commission as a naval engineering officer. The Guard has paid for my additional training since then, including a master's degree, but all new accessions have to have a bachelor's degree in naval or maritime engineering first before they can put in a package for a commission. I took advantage of the Coast Guard Student Pre-Commissioning Program to go to college. I had to go to both enlisted boot camp, which was hard, and Officer Candidate School, which was harder, but my last two years of college were more interesting than they ever could have been if I had not joined the program. The scholarship was helpful, too.

Coast Guard cutters are built at the same shipyards that build naval vessels. In fact, the six large shipyards remaining in the United States get most of their work from Navy and Coast Guard contracts. There is very little shipbuilding left in the United States, but at least what is still done is the most advanced in the world from a technology and engineering standpoint.

Cutters are designed by teams of engineers composed of Coast Guard naval engineers and naval engineers from the companies that actually build the vessels. We are the experts in what the service needs its ships to do, and they are the experts in construction techniques. Between us, we cover all the angles. Today's cutters are extremely sophisticated pieces of machinery. They involve the work of many different contractors, too. The shipyard that builds the actual hull is usually the prime contractor, but there can be hundreds of subcontractors building various pieces and subsystems that make the ship complete. I get to collaborate with all of them, which is perfect for someone with my interests.

Our big project right now is called Deepwater. It is a very ambitious project to update all of the Coast Guard ships and aircraft in a comprehensive, thought-out way. All electronic systems, for example, have to be compatible with all the others and with those used by the Navy, our sister service. This is a much more complicated job than it may seem, and finishing will likely take a decade or two longer than anybody would like. Ships are very expensive and are easy to cross out of a budget. If somebody in Congress wants to free up a billion dollars all they have to do is get a single high-endurance cutter cancelled.

I recommend this career to people who are really passionate about ships and shipbuilding. We are operating at the cutting edge. We do scientific research that has led to remarkable advances in ship design. This is not a career for people who want to go along to get along. You have to be at the top of your game every day or you will fall behind."

I Am a Damage Controlman

"I have one of those dirty jobs nobody wants to do. I prepare for the unthinkable. I am a damage controlman. It is my job to make sure that our cutters always sail with the people and materials necessary to keep them afloat if they are damaged, whether in combat or due to an accident or fire.

I picked this rating after boot camp. Everybody does basic damage control training when they join the Coast Guard. The philosophy is simple. It doesn't matter if you are a seaman or an admiral. If the ship goes down, you go down with it. In the event of an actual emergency aboard ship, everybody has a role to play. Everybody does damage control in an emergency. Damage controlmen are the specialists for everyday work.

I spend most of my time doing training. When we get under way my team of damage controlmen starts at one end of the ship and works toward the other. We run damage control exercises for the rest of the crew, keeping their skills up to par. When we get to the far end of the ship we do it all over again. We also run large-scale exercises for the entire crew. In the event of an emergency we will be in charge of damage control efforts, but we will need the help of everyone on board. It works much better if everybody knows what they are doing.

A ship damaged in combat is just a sitting duck. If we can't return fire we will be destroyed. That is why Navy and Coast Guard vessels have redundant systems. That is, if one engine gets knocked out there is a second one to take over. If one firefighting system goes down there is a backup. This is really important to keeping the ship in the fight.

You might be surprised at some of the things we use in our job. We sometimes plug holes with corks. Yes, you read that correctly – corks. Really big corks. We also use simple box patches, which are just wooden boxes shored up with heavy timbers or jacks. We have even been trained in how to stuff a mattress into a hole. It works. It's not our job to make permanent repairs. It's our job to keep the ship in the fight.

I don't know if I'll stay in for 20 years. I've only been in for three years. I joined because I wanted to get the GI Bill so I can go to college. The Coast Guard also sounded exciting, and it definitely is. In fact, I didn't know it was going to be this exciting. I thought I'd stay in for one hitch and then go to college. Now I'm thinking I'll stay in and go to college part time. I can't think of anything I could do in civilian life that would be as rewarding as the work I am doing now."

I Am an Intelligence Specialist

"I can't tell you what I do. Well, I can tell you what I do, but I can't tell you the specifics. I am an intelligence specialist.

When they hear the word 'intelligence' many people think of secret agents. Actually, intelligence is whatever information the commander needs to make better decisions. This can be anything from classified intelligence originating with a human source to something as simple as a weather report.

This is a new rating for the Coast Guard. The other services have had intelligence specialists for decades. The Coast Guard covered most of the bases covered by today's intelligence specialists but they did it in a sort of roundabout way, with pieces of the larger intelligence picture being handled by various specialties throughout the service. The change was made in the wake of the September 11, 2001 terrorist attacks on the United States. The Coast Guard dramatically stepped up its responsibilities for boarding and searching vessels entering U.S. ports. Legally, we are entitled to inspect any ship headed for a US port. We are always required to know as much as possible about them, even if we choose not to board them.

This created a huge demand for people specializing in gathering and disseminating intelligence. We need to keep track of all the ships that enter our waters. We need to know where they're coming from, where they've been, who is on their crew, who they're owned by, what they're carrying, what they've carried in the past, where they're going next, who their insurance company is, and what country they're registered in. Based on those data we have to decide which vessels are what we call Contacts of Interest, or COIs. COIs are tracked and boarded at the first opportunity. Intelligence specialists make this recommendation and then brief boarding teams before they go. We tell them what they should be looking for. A COI may be able to hide contraband or damning evidence the first time we board the ship. When the boarding team returns we get to work analyzing whatever they bring us. Copies of crew lists, cargo manifests, cell phone call logs and ship's logs. Usually we do not make any arrests, but we always learn something to help us plan for future efforts.

We have also applied this expertise to assisting the Navy with patrolling some of the nastier corners of the world. The Navy has its own boarding teams, but we supplement them when we can and conduct law enforcement boardings that the Navy is prohibited by law from conducting. Remember, only the Coast Guard can do law enforcement as well as military operations.

I would recommend this career to anybody with a good analytical mind and endless patience. If we have a failing in the military intelligence community, it's that we have too much information to analyze and not enough analysts to do it. Nine times out of 10, when something falls through the cracks it is not because we did not know about it, it's because we didn't put all the pieces together. That is why I have excellent job security. My work is needed and will continue to be."

PERSONAL QUALIFICATIONS

ALTHOUGH MOST PEOPLE ADMIRE THE COAST GUARD AND THE WORK THAT IT DOES, MOST people would not last very long as Coast Guardsmen. To succeed you must be adaptable, physically fit and dedicated to serving others.

More than any other military service, the Coast Guard requires its members to step up and take responsibility for decisions and actions not exactly in their job descriptions. This is doubly true for officers, who are almost always required to take a few jobs outside their specialty during their careers. Enlisted personnel specialize in their jobs, but the rough-and-tumble nature of Coast Guard duty means that everybody has to be willing to lend a hand. Coasties have a very cohesive, supportive culture that does not allow for complaining. If a shipmate asks you to grab the other end, you grab the other end. "That's not my job!" is not an acceptable response. As your career progresses you will find that you are better at some things than you are at others. The service will respond accordingly and you will probably be assigned to duties that will be a good fit for you. The Coast Guard does not want to assign people to jobs they do not like and are not good at. Still, you will be happier in the Coast Guard if your idea of a new challenge and your idea of a good time are the same thing.

All the military services maintain high physical fitness standards. Some, like those for Special Forces branches, are almost superhuman. Most, however, are not. They are challenging, however, and will require sustained effort on your part. The Coast Guard maintains a culture of fitness that you should plan to fit into. Coasties don't pump iron the way Marines do. They do, however, do serious calisthenics, run, and swim, swim, swim. Even if your job keeps you behind a desk most of the time you will be expected to be a good swimmer. No matter what your specialty, you will get underway aboard cutters from time to time. Accidents happen. People fall overboard. You are no good to the team if you are the one who needs to be rescued. Get fit, stay fit and go to the pool regularly.

If you join the Coast Guard, you will be serving the public in a very meaningful way. A very visible way, too. Coast Guard rescues almost always get news coverage, and in survey after survey, Americans say they admire the Coast Guard because its members do such amazing things to help people. This service is its own reward. If you believe this, there may be a place for you in the Coast Guard.

ATTRACTIVE FEATURES

THE COAST GUARD OFFERS MANY attractive features for potential careerists. Coasties see more action than most military personnel do. "Seeing some action" has been the goal of military personnel for centuries. After training, training and more training, people in the military look forward to the day they get to put their skills to the test. Nobody wants to go to war, but everybody wants to see their efforts lead to something tangible. Most military personnel go through years of training that may be punctuated by short stints in combat zones. Even there, relatively few personnel ever get to the front lines. Service in the Coast Guard is different. Like a police department, the Coast Guard is involved in its own low-level war 365 days a year. If you pursue a career in the Coast Guard, you will participate in law enforcement actions and at-sea rescues as soon as you finish your initial training. And unlike war, which can be morally and politically complicated, fighting crime and rescuing people will make you feel good every single time. You will be doing one of those hard jobs somebody has to do and you will have every right to take pride in yourself.

Like the other military services, the Coast Guard is a bureaucratic organization and prone to the frustrations and inefficiencies that plague all bureaucracies. Unlike the other services, however, the Coast Guard is small enough to enable your voice to be heard. Many Coast Guard installations only have a few dozen personnel assigned to them, which makes it much easier to get things done. The Guard's unique culture of expecting everybody to do everything also helps to move things along with fewer delays and less paperwork. In this regard, the Coast Guard's small size is definitely an asset.

There is certainly nothing wrong with a career where you are universally admired. The Coast Guard does good work and everybody appreciates it. The other armed services will never have the degree of positive public approval afforded the Coast Guard because the other services have to bear the burden of being involved in military actions that some people do not approve of. The Coast Guard never injures innocent civilians when a bomb goes astray, never has to deal with TV news showing soldiers kicking down the wrong door, and rarely flexes its muscle in places few Americans have heard of or care about. When the Coast Guard engages in military campaigns it does so in a supporting role, usually conducting the same search-and-rescue and law enforcement missions it does at home.

UNATTRACTIVE FEATURES

WORK IN THE COAST GUARD IS MADE harder by the fact that the service is always overstretched and underfunded, even with the new emphasis on solving those problems.

Most people enjoy being challenged. Most people do not like to be pushed to the limit day after day. If you pursue a career in the Coast Guard you will have tours of duty that require you to do your utmost each and every day. When you are assigned to sea duty, for example, you could spend weeks or months at a time at sea, living aboard a ship and participating in high-intensity activities. When the weather gets rough the Navy just sails away from the storm. The Coast Guard's job, however, is to sail *into* the storm to rescue people in danger. Even if your job takes place entirely within the skin of the ship and does not directly expose you to the hazards of open water you will still be expected to sail into harm's way with the rest of the crew and continue to do your job even while the ship is pitching and most sailors would call it a day. Not every tour will be this strenuous, but you will be expected to be at the peak of your abilities, with no slow days or excuses, for very long periods of time.

This situation is exacerbated by the fact that the Coast Guard has been chronically underfunded for so long that its ships and aircraft are not as well equipped as they should be. Although much smaller than the US Navy, the Coast Guard is actually one of the world's largest maritime forces. Its fleet, however, is one of the oldest in the world. The Navy, for example, would never keep a ship in commission for more than 50 years, but the Coast Guard fleet includes several cutters much older than anybody aboard them. The Deepwater program and the new emphasis on the Coast Guard's unique mission have alleviated this to some degree, but the Coast Guard still receives a fraction of the money per person that the Navy receives. You will find that you are expected to do more with less, all the time.

If you pursue a career in the military or law enforcement you can never, ever forget that you are taking a pledge to lay your life on the line if that is what it takes to protect the people you serve. That is a serious commitment. All police officers and military personnel hope they never have to face such a situation. The fact remains that they have to be prepared to make the ultimate sacrifice if the need arises. This is not something to be taken lightly or with an attitude of bravado.

EDUCATION AND TRAINING

BEYOND A HIGH SCHOOL DIPLOMA there are no formal educational requirements to join the Coast Guard. As with most other professions, however, a career in the Coast Guard will go farther if you complete additional education and training as your career progresses. This is demanding work that requires you to think on your feet, sometimes under very demanding circumstances. The more you know, the better you'll do.

If you plan to enlist in the Coast Guard, the first thing you need to do is graduate from high school. All of your subjects are important. Mathematics will help you to unravel the secrets of navigation. Physics will teach you about applied sciences like naval engineering and the challenges inherent in building and maintaining infrastructure in and near the water. Good English skills will help you to communicate and to organize your thoughts clearly. Any foreign-language skill will be put to use by the Coast Guard, and do not forget gym class! Physical fitness is a basic necessity of the Coast Guard culture. You must be fit to take on physically demanding duties like serving as a rescue swimmer. Just serving aboard a ship at sea is demanding, and you must be in good shape.

You will do plenty of workouts! Coast Guard boot camp consists of eight weeks of rigorous training in Cape May, New Jersey. You will spend part of each day in classroom training, learning about Coast Guard history, maritime law and how the service bureaucracy works. You will also spend part of each day doing physical training, or PT. The Coast Guard guarantees that you will leave boot camp in the best shape of your life, and they mean it. Most days you will also get to tackle practical exercises like seamanship training, firefighting and damage control. The first week or so can be intimidating, but you will enjoy boot camp once you settle into a rhythm.

If you want to apply for a commission you have several options. The Coast Guard College Student Pre-Commissioning Initiative is similar to Reserve Officer Training Corps, or ROTC, in the other services. The CSPI program is an excellent option.

Many Coast Guard officers got their start by applying for a commission after graduating from a civilian college or university. A bachelor's degree will get you started. If you have a degree, or are approaching your last semester of college, sit for an interview with an officer recruiter. If the recruiter thinks you have potential you will be invited to prepare an application package. A complex and time-consuming process, the application package must be completed with great precision. Any deviation from the instructions could derail the entire application. Take your time and do not be afraid to ask for input from the recruiter. You will also have to sit for an interview with a panel of Coast Guard officers. If you complete this process successfully, you will have the opportunity to attend Officer Candidate School.

Officer Candidate School (OCS) consists of 17 weeks of training at the Coast Guard Academy in New London, Connecticut. Training is rigorous and broad, as Coast Guard officers are expected to be generalists who can do whatever the service needs them to do. Initial assignments are made during OCS. Officer candidates are asked their preferences, but initial assignments are always made according to the needs of the Coast Guard. Your personal desires will be considered but they will never be the deciding factor. OCS is similar to boot camp in that your days will be filled with a combination of PT, classroom teaching and hands-on technical training. When you graduate you will be commissioned as an ensign and obligated to serve for three years.

If you are especially ambitious you can apply to the Coast Guard Academy. Much smaller than the other service academies, the Coast Guard Academy is an undergraduate college that accepts only about 300 of the high school graduates who apply each year. Unlike the other service academies, the Coast Guard Academy does not require applicants to secure a nomination from a member of Congress to be considered. Academic standards are every bit as high, however, and only a small fraction of applicants are accepted. Applicants are judged on their academic performance, leadership potential, physical fitness and civic involvement. If you want to go to the academy, you really need to get started now. If you are accepted, there is no tuition and the Coast Guard will actually pay you a small salary during your four years in attendance. You can choose from among eight majors, all of which lead to a bachelor's degree. When you graduate you will be commissioned as an ensign and be required to serve for at least five years.

No matter which route you take to a Coast Guard career do not let an irrational fear of initial access training put you off. Depicted in movies as a demeaning process filled with screaming drill sergeants and daily doses of humility, boot camp and OCS are largely academic exercises today. You will spend most of your time in classrooms. PT standards are challenging, but they are not superhuman. The training can indeed be humbling, but for good reason. You are about to be entrusted with much more power and authority than most young people can even dream about. You have to prove that you are worthy of the enormous trust placed in you by the Coast Guard and the people it serves. Folding your underwear a certain way may seem silly but there is logic to it: If you can't be trusted to take an order to fold your underwear properly why should you be trusted to command a ship? Boot camp and OCS are not supposed to be easy. After the first week most people start to like the experience.

You will continue to spend some of your time in classrooms as your career progresses. In fact, when you are not actually involved in real-world operations, you will be in some sort of training. Performing rescues at sea is very dangerous and something that needs to be rehearsed regularly. The same goes for boarding suspected criminal vessels or assisting the Navy with search-and-seizure operations in foreign waters. The Coast Guard's business is too risky to leave anything to chance.

EARNINGS

NOBODY GETS RICH IN THE MILITARY but everybody does pretty well. The compensation package is designed to make service members feel secure in the knowledge that they, their families and their futures are being looked after. It is very difficult to get people to go to sea for months at a time if they feel their loved ones will not be able to get by.

In all military services your paycheck is actually composed of several different kinds of compensation. The largest portion of your paycheck is base pay. This is your salary without benefits. In 2009 a new enlisted Coastie in the rank of E-1 (seaman recruit) was paid $1,400 per month, or $16,800 per year. This may not sound like much, but keep in mind that most E-1s are in boot camp where they also get a roof over their head, three meals a day and almost no opportunities to spend money. Base pay jumps to $1,569 per month for E-2s (seaman apprentices) and $1,588 for E-3s (seamen).

As a young enlisted Coastie or officer, most of your needs will be provided for, so your modest paycheck will actually go pretty far. When you rise to E-4 (petty officer third class) your base pay will jump to $1,921 per month if you have been in for at least two years, or $2,025 per month if you've been in for three years. An E-5 (petty officer second class) with four years in earns $2,335 per month. After that, your base pay will increase every two years all the way up to $5,928 per month for an E-9 (master chief petty officer) with at least 26 years of service. The highest-ranking personnel can, under certain circumstances, serve for as long as 40 years. With a combination of hard work and luck, you could top out the pay scale at

$6,830 per month in base pay for an E-9 with at least 38 years of service.

At some point you will probably move off base and into a house or apartment of your own. When you do you will be entitled to a housing allowance. Housing allowances are calculated based on your rank, location and whether or not you have dependents. The housing allowance for a single E-5 living in a rural area, for example, would be substantially less than the housing allowance for a married E-5 living in an expensive metropolitan area. Housing allowances for some areas can be as large as or larger than your base pay. Typically, however, a housing allowance will account for about one third of your paycheck.

After your base pay and housing allowance, there are additional entitlements. The cost of living allowance, or COLA, adds to your base pay when you are assigned to an area with a high cost of living. Hazardous duty pay means more money when you deploy to a dangerous area. Your paycheck will be tax free if you are deployed to an area declared a combat zone by Congress. Personnel who can prove their fluency in certain languages may be eligible for a monthly bonus. The list of ways to boost your paycheck is quite lengthy. All of these bonuses will require hard work on your part.

Pay scales for officers are higher than those for enlisted personnel. Most other allowances are higher too. Base pay for a new O-1 (ensign) starts at $2,655 per month and rises to $3,483 for an O-2 (lieutenant junior grade) with two years of service. From there the pay scale rises to a maximum of $18,061 per month for an O-10 (admiral) with at least 38 years of service. There is only one four-star admiral serving at any given time: the commandant.

The Coast Guard also fields about 400 warrant officers, an intermediate rank that can only be attained after you have achieved the rank of petty officer first class. Nobody becomes a warrant officer by being promoted. If you want to be a warrant officer you will have to apply for a position. Warrant officers are technical experts who devote their careers to a specific technical specialty. In the chain of command they are above all enlisted personnel but below all commissioned officers. Not all designators have warrant career paths. Base pay starts at $3,871 per month for a CW-2 (the CW-1 rank is no longer in use) with at least 8 years of service, the minimum time in service needed to apply for the warrant program. A CW-3, or chief warrant officer three, the next step in the chain, earns $4,677 per month with at least 14 years of service. The warrant ranks top out at CW-4. With 38 years of service base pay maxes out at $6,815.40 per month.

OPPORTUNITIES

YOUR COAST GUARD CAREER CAN START sooner than you think.

All of the military services pride themselves on their all-volunteer forces. The Coast Guard takes this idea one step further with the United States Coast Guard Auxiliary, an organization of 30,000 serious, hard-working, unpaid volunteers who give their time to support Coast Guard missions. Auxiliarists, as they call themselves, come from all walks of life and include everybody from the very young to the very old. They support Coast Guard missions in many different ways, including maintaining aids to navigation, using their own boats and aircraft to conduct surface patrols, and conducting boating safety classes and inspections at recreational marinas across the country. Auxiliarists can take advantage of a surprisingly large array of training opportunities and can learn new skills to support the mission and enhance their own proficiency with boats and navigation. Auxiliarists are also heavily involved in public affairs, representing the Coast Guard in small communities across the country. Auxiliarists get impressive uniforms, rank insignia – which is technically honorary but generally respected by others – and opportunities to earn medals and ribbons just like their active-duty counterparts. You should check out opportunities with an Auxiliary in your area.

The Coast Guard does not maintain a Reserve Officer Training Corps program like the other services, so conventional ROTC is not an option for aspiring Coast Guard officers. The service instead offers the Coast Guard College Student Pre-Commissioning Initiative, a program that comes with benefits similar to those offered by ROTC programs. College students who want to earn commissions in the Coast Guard can apply for the CSPI in their sophomore year. If accepted they receive base pay at the E-3 pay grade and a college scholarship. There are a few drawbacks, however, because CSPI students can go to college anywhere and do not necessarily have other CSPI students around them to form a conventional ROTC unit. CSPI students have to attend enlisted boot camp before they start their junior year of college. During their junior and senior years, CSPI students must attend Coast Guard training events from time to time, similar to an ROTC program. Students have to attend Officer Candidate School upon graduation. CSPI graduates are required to serve for a minimum of three years after completing OCS. If you take this route, you will be guaranteed admission to OCS. The experience will likely put you at the head of your class, too. The mandatory three-year commitment is shorter than the requirement imposed by the other services, all of which require four or five years of service. t one.

GETTING STARTED

YOUR NEXT STEP SHOULD BE TO TALK TO a recruiter. The Coast Guard's needs change constantly, and only a recruiter can give you timely information you will need to get your career started.

To enlist you will need to take the Armed Services Vocational Aptitude Battery, or ASVAB, test before you can negotiate a contract. You will need to achieve a minimum score of 36 to get into the Coast Guard, although many rates require higher scores. Taking the ASVAB does not obligate you in any way. If you decide you do not want to join the Coast Guard you can walk away or use your score to enlist in one of the other services. Taking the ASVAB is the first exposure many people have to an actual military environment. Typically, you will take the ASVAB at your nearest Military Entrance Processing Station, or MEPS. After you get the results you can start to negotiate a contract. Pick your rate, spell out which schools you will go to to start your career and determine when you will report to boot camp. Your obligation begins as soon as you sign the contract and take the oath.

Officer candidates intending to go to Officer Candidate School must also take the ASVAB. Generally you will do this either immediately after you finish college or during your last semester. You must already possess a bachelor's degree in order to be commissioned. You can start the commissioning process when you are still in college but everything will be provisional until you graduate. You will generally be able to negotiate your initial training plans before you sign a contract. This is especially true for officer candidates with degrees in technical disciplines like engineering.

As soon as the ink has dried on your contract you will be in the Coast Guard's hands. Luckily, that's a good place to be. You will acquire skills and experience you cannot get anywhere else. Whether you stay in for one hitch or go for a 20-year career, your life will be more interesting and rewarding for your time spent in the Coast Guard. Good luck!

ASSOCIATIONS
PERIODICALS
WEBSITES

- **Africa Partnership Station**
  www.c6f.navy.mil

- **African Center for Strategic Studies**
  www.africacenter.org

- **American Legion**
  www.legion.org

- **Canadian Coast Guard**
  www.ccg-gcc.gc.ca/eng/CCG/Home

- **Canadian Navy**
  www.navy.forces.gc.ca

- **China Defense Today**
  www.sinodefence.com

- **Combined Joint Task Force Horn of Africa**
  www.hoa.africom.mil

- **Defense Link**
  www.defenselink.mil

- **Defense News**
  www.defensenews.com

- **Foundation for Coast Guard History**
  www.fcgh.org

- **French Foreign Legion**
  www.legion-recrute.com/en/

- **French Navy**
  www.defense.gouv.fr/marine_uk

- **Global Security**
  www.globalsecurity.org

- **Guardia Costiera d'Italia**
  www.guardiacostiera.it/en

- **Haze Gray and Underway**
  www.hazegray.org

- **Her Majesty's Coastguard**
  www.mcga.gov.uk/c4mca
  /mcga07-home

- **Indian Coast Guard**
  www.indiancoastguard.nic.in

- **Indian Navy**
  www.indiannavy.nic.in

- **International Journal of Naval History**
  www.ijnhonline.org

- **Irish Naval Service**
  www.military.ie/naval

- **Israeli Defense Forces**
  www.dover.idf.il/IDF/English

- **Jack's Joint**
  www.jacksjoint.com

- **Jane's**
  www.janes.com

- **Japanese Maritime Self-Defense Force**
  www.mod.go.jp/msdf/formal
  /english/index.html

- **Marshall Center for European Center for Security Studies**
  www.marshallcenter.org

- **Military.com**
  www.military.com

- **Naval Open Source Intelligence**
  www.nosi.org

- **Naval Postgraduate School**
  www.nps.edu

- **Naval Review**
  www.naval-review.org

- **Naval Technology**
  www.naval-technology.com

- **Naval War College Review**
  www.nwc.navy.mil/press
  /review/overview.aspx

- **NavSource**
  www.navsource.org

- **Navy League of the United States**
  www.navyleague.org

- **Navy Times**
  www.navytimes.com

- **North Atlantic Treaty Organization**
  www.nato.int

- **Royal Australian Navy**
  www.navy.gov.au

- **Royal National Lifeboat Institution**
  www.rnli.org.uk

- **Royal Navy**
  www.royalnavy.mod.uk

- **Royal New Zealand Coast Guard**
  www.nzcoastguard.org.nz

- **Royal New Zealand Navy**
  www.navy.mil.nz

- **Russian Navy (unofficial)**
  www.rusnavy.com

- **Safe Boating America**
  www.safeboatingamerica.com

- **Sea Waves Magazine**
  www.seawaves.com

- **Sea Rescue**
  www.sea-rescue.de/english

- **Semper Paratus**
  www.semperparatus.com

- **Today's Military**
  www.todaysmilitary.com

- **United Nations**
  www.un.org

- **United Nations Security Council**
  www.un.org/docs/sc

- **United States Air Force**
  www.af.mil

- **United States Army**
  www.army.mil

- **United States Coast Guard**
  www.uscg.mil

- **United States Coast Guard Academy**
  www.cga.edu

- **United States Coast Guard Auxiliary**
  www.cgaux.org

- **United States Department of Defense**
  www.defenselink.mil

- **United States Department of Homeland Security**
  www.dhs.gov

- **United States Department of Veterans Affairs**
  www.va.gov

- **United States Institute of Peace**
  www.usip.org

- **United States Marine Corps**
  www.marines.mil

- **United States Naval Institute**
  www.usni.org

- **United States Navy**
  www.navy.mil

- **United States Power Squadrons**
  www.usps.org

- **Veterans of Foreign Wars**
  www.vfw.org

- **Warship International**
  www.warship.org